College Dormitory Fires in Dover, Delaware and Farmville, Virginia

Investigated by: Daniel J. Carpenter, Jr.

This is Report 006 of the Major Fires Investigation Project conducted by TriData Corporation under contract EMW-86-C-2277 to the United States Fire Administration, Federal Emergency Management Agency.

 **Homeland Security**

Department of Homeland Security
United States Fire Administration
National Fire Data Center

U.S. Fire Administration Fire Investigations Program

The U.S. Fire Administration develops reports on selected major fires throughout the country. The fires usually involve multiple deaths or a large loss of property. But the primary criterion for deciding to do a report is whether it will result in significant "lessons learned." In some cases these lessons bring to light new knowledge about fire--the effect of building construction or contents, human behavior in fire, etc. In other cases, the lessons are not new but are serious enough to highlight once again, with yet another fire tragedy report. In some cases, special reports are developed to discuss events, drills, or new technologies which are of interest to the fire service.

The reports are sent to fire magazines and are distributed at National and Regional fire meetings. The International Association of Fire Chiefs assists the USFA in disseminating the findings throughout the fire service. On a continuing basis the reports are available on request from the USFA; announcements of their availability are published widely in fire journals and newsletters.

This body of work provides detailed information on the nature of the fire problem for policymakers who must decide on allocations of resources between fire and other pressing problems, and within the fire service to improve codes and code enforcement, training, public fire education, building technology, and other related areas.

The Fire Administration, which has no regulatory authority, sends an experienced fire investigator into a community after a major incident only after having conferred with the local fire authorities to insure that the assistance and presence of the USFA would be supportive and would in no way interfere with any review of the incident they are themselves conducting. The intent is not to arrive during the event or even immediately after, but rather after the dust settles, so that a complete and objective review of all the important aspects of the incident can be made. Local authorities review the USFA's report while it is in draft. The USFA investigator or team is available to local authorities should they wish to request technical assistance for their own investigation.

For additional copies of this report write to the U.S. Fire Administration, 16825 South Seton Avenue, Emmitsburg, Maryland 21727. The report is available on the Administration's Web site at http://www.usfa.dhs.gov/

U.S. Fire Administration

Mission Statement

As an entity of the Department of Homeland Security, the mission of the USFA is to reduce life and economic losses due to fire and related emergencies, through leadership, advocacy, coordination, and support. We serve the Nation independently, in coordination with other Federal agencies, and in partnership with fire protection and emergency service communities. With a commitment to excellence, we provide public education, training, technology, and data initiatives.

TABLE OF CONTENTS

COLLEGE DORMITORY FIRES IN DOVER, DELAWARE AND FARMVILLE, VIRGINIA

OVERVIEW

The potential threat of college dormitory fires is often not taken seriously enough by students until it is too late. Campus authorities and students sometimes let their guard down because of the high frequency of pranks and false alarms.

On April 12, 1987, in Williams Hall of Wesley College, Dover, Delaware, and on April 28, 1987, in Frazer Dormitory of Longwood College, Farmville, Virginia, fires occurred that killed one student and injured nineteen. While the fires differ, they have several important aspects in common.

Smoke bombs caused the Wesley College, Delaware fire in which an 18-year-old student died and four students were injured, one critically. Incidents involving smoke bombs had occurred before, and students apparently thought the smoke this time "was just another smoke bomb." As a result the fire department was not immediately notified.

There had been frequent false alarms, and students considered them annoying. The fire alarm in the dormitory did not operate on the day of the tragic fire, apparently because the fire alarm bell had been stolen from the first-floor hallway after a false alarm the previous day.

When someone did call for help, the call went to the Dover police department on its regular phone number, instead of to the fire department via the 911 emergency number, further delaying the response.

Delayed notification of the fire department also characterized the Longwood College, Virginia, dormitory fire. The call to the Farmville Volunteer Fire Department was relayed through the campus police department; the 911 emergency number was not used.

The Longwood College fire, which was apparently caused by an unauthorized overloaded, multi-outlet extension cord, injured 15 students. As in the Wesley College incident, the fire alarm failed to work. In this case, activation was delayed about 10 minutes because the breaker switch was off. In addition, the majority of in-room smoke detectors were disconnected or failed to operate.

Similar to those at Wesley College, the Longwood College students apparently did not evacuate immediately because they thought it was "just another fire drill."

Both these incidents point out the importance of enforcing fire safety policies and procedures in dormitories and encouraging use of the 911 emergency number for reporting emergencies to appropriate authorities.

Gathering information for these reports was somewhat restricted by the criminal and civil investigations at Wesley College and the lack of incident reports and interviews with the fire department in Farmville. Nevertheless, the following summary of problems emerged:

SUMMARY OF KEY ISSUES

Issues	Longwood College Farmville, Virginia	Wesley College Dover, Delaware	Comments
Cause of Fire	Overloaded circuit	Smokebomb prank	Both avoidable by proper student behavior.
Smoke Detectors:			
Room of origin	Present but did not work	No detectors	Lack of maintenance and vandalism by students (Longwood).
Other rooms	Present; about 85 percent inoperative	No detectors	Should be operational in all dorm rooms.
Alarm System	Failed	Failed	One vandalized, the other accidentally turned off.
Sprinklers	None	None	Dormitories built prior to code requirement for sprinklers.
Evacuation	Delayed	Delayed	Lack of student belief of a real fire.
F.D. Notification	Delayed	Delayed	Calls went to local or campus police before fire department. 9-1-1 number not used.
Firefighting Problems	First responder (campus police) lacked air pacs, other needed equipment and training. First responding engine company had empty air pacs.	First responder misdirected to wrong side of building.	Interior attack delayed.
Structural Problems	Lack of self-closing room doors.	No problem	Probably contributed to smoke spread (Longwood).

WESLEY COLLEGE DORMITORY FIRE
Dover, Delaware, April 12, 1987

Local Contacts:

Police Department
Capt. James L. Hutchison (302) 736-7010
Lt. Boyce Failing
400 South Queens Street
Dover, Delaware

Fire Department
Chief John Wilson (302) 734-3302
Bureau of Fire
Robbins Hose Company
103 South Governors Avenue
Dover, Delaware 19901

Inspector John W. Raughley (302) 736-7010
Department of Inspections
P.O. Box 475 – City Hall Annex
Dover, Delaware 19903

Walter E. Saxton (302) 736-7010
Inspections Supervisor
P.O. Box 475 – City Hall Annex
Dover, Delaware 19903

Wesley College
Mr. Joseph Bellmyer (302) 736-2360
Assistant to the President
Wesley College
Dover, Delaware 19901

Dr. J. Thomas Sturgis (302) 736-2435
Dean of Students
Wesley College
Dover, Delaware 19901

OVERVIEW

A tragic dormitory fire in Williams Hall at Wesley College in Dover, Delaware, occurred in the early morning hours of Sunday, April 12, 1987. It took the life of one student, critically injured another (who subsequently recovered), caused lesser injuries to three more students, and brought fear and panic to dozens of others.

The fire started when a smoke bomb was set off in one of the rooms. This was the tragic climax of a series of similar incidents in the dormitory. Evidence of previous smoke bombs which burned hall tiles was apparent throughout the building. Lighter fluids had been sprayed on the wall and ignited, and smoke from burning candles had been used to write obscene words on the ceilings. In addition, false alarms had been frequent prior to this incident, and fire alarms were regarded only as an annoyance by many students.

CONSTRUCTION

Wesley College was founded in 1873. It is located in Dover, Delaware, in a scenic rural setting. There are approximately 800 full-time students and 700 part-time day and evening students enrolled.

The Williams Hall dormitory was built in 1969. It is a three-story fire resistant building with brick exterior and concrete block interior, and it has a basement. Two-inch solid wood, self-closing doors are at stairways located at opposite ends of the building. Student rooms have a two-inch solid wood door with self-locking door knobs. The dormitory is designed to house approximately 180 students. It had no sprinklers or smoke detectors in each room.

Each of the college's five dormitories are equipped with a local fire alarm system on every floor. However, a fire alarm bell had apparently been removed from the alarm in the first floor hallway of Williams Hall following a false alarm during a party the morning prior to the fire, April 11. (Fire alarm bells had previously been found missing and the alarms inoperative in a fire drill in September, 1986).

According to the fire officials contacted, the tampering with the alarms may have caused the alarm system to be inoperable at the time of the fire on the morning of April 12, 1987. School officials agreed, but stated that it was not possible to replace the alarm bell immediately since the original could not be found, and because it was a weekend and the alarm system company was not open. There were several other reasons as well. However, no special fireguard or other actions were taken as an interim measure.

Panic hardware equipped with an alarm device on the exit doors leading outside from the basement did function properly when students exited the building through these doors on the morning of the fire. It is also believed that the alarm system in an adjacent dormitory may have been activated in an attempt to alert students.

POLICIES AND PROCEDURES

Information about fire safety regulations and procedures was readily provided to the investigator by Wesley College. This included a schedule of fire drills, fire safety precautions to be observed by students, and procedures to be followed in the event of a fire. False alarms are counted toward meeting the number of fire drills required per month.

A residence director is assigned to each dormitory and is assisted by two other resident assistants (R.A.s) on each floor. At no time are the resident director or assistants permitted to leave the building unattended. The name of the duty R.A. is posted to insure that someone in authority is available in the event of an emergency. There were four R.A.s in the dorm at the time of the fire.

While it appears that the policies and rules of conduct were in-place to insure a reasonably fire safe environment, it is also apparent that these rules were not fully enforced. Interviews with college and fire officials indicated a supreme concern for what was regarded as each student's individual rights. The dormitory was considered to be a student's home, and officials seemed reluctant to enforce rules which would put the need of the total resident population ahead of the "rights" of the individual to determine his/her own behavior within his/her home.

THE FIRE

On the morning of February 12, 1987, a telephone call was received at approximately 0233 on the house phone by the radio dispatcher at the Robbins Hose Company #1, advising the department of a fire in the Williams Hall dormitory at Wesley College. The call was relayed by an answering service which takes calls for the department after hours.

The Robbins Hose Company #1 is a well-established volunteer fire department supported by a paid staff which includes dispatchers, fire prevention inspectors, and investigators.

Before the fire department had been called directly, notification of a fire had been transmitted to the Dover Police Department dispatcher who also relayed it to the fire department. The call had come in on regular police lines rather than the 911 emergency number. According to the police, this was due to the fact that the 911 emergency number had only been in effect for approximately two years, and "some people are accustomed to just dialing the police department on the regular police phone number."

Fire officials determined later that the calls to the police and fire departments had been delayed for some time. Initially, building occupants did not realize there was a fire even though there was smoke. Most believed it was "just another smoke bomb being set off."

Once an actual fire was discovered, students said they attempted to alert everyone in the building, but because of self-locking doors on the individual rooms, they were unable to determine which of the rooms were occupied.

Because of the magnitude of the fire, and the possible extent of rescue which could be involved, a second alarm was called soon after arrival of the first unit. A total of 10 units and 41 firefighters responded.

Ironically, all dormitories except Williams Hall had been pre-fire planned. Because of the similarity among the dorms, the pre-fire plans for the other buildings assisted firefighters in their rescue attempts and firefighting operations. However, a further delay was caused by misinformation received by the first arriving units at the scene. They were directed to the rear of the building, from which smoke could be seen, but the actual area of origin turned out to be at the front of the building.

The body of 18-year-old Christopher T. Steven was found in Room 220 during the firefighters' initial search of the building. He had died of smoke inhalation. Later, at approximately 0330, firefighters discovered a second victim who was still alive, located in Room 217. Both of these victims were at the opposite end of the dormitory from the room of origin, a distance of approximately 175 feet. The second (alive) victim was covered with a blanket and found hidden under a desk. When questioned by investigators, he stated that he had attempted to leave the building but became confused and disoriented. He, therefore, returned to his room, closed the door, and hid under the desk in an attempt to avoid the fire. Because of his concealment, he had not been discovered during the initial search by the firefighters and was not discovered until almost an hour after the fire was reported to the fire department.

Damage from the fire is estimated at more than $100,000.

LESSONS LEARNED AND RECOMMENDATIONS

1. Vandalism of fire safety systems and pranks involving the setting of fires simply cannot be tolerated by colleges. Students need to be made to understand the dangers from fire, and the school and criminal penalties they may face for mischief.

    Concern for students' individual rights to privacy and self-expression is fine, but there must also be recognition and protection of the right of all students who share a building to live in a fire safe environment. Student orientation programs should spell out strict disciplinary measures to be taken when infractions of college safety procedures occur, and they should be enforced consistently.

2. The advantages of the 911 reporting procedures should be thoroughly explained and utilized. Any delay when an emergency is known to exist should be thoroughly investigated to determine cause and the necessary action to be taken to avoid the problem in the future.

3. Precautions regarding criminal and civil investigative procedures are important in cases such as this. Two students who lived in Williams Hall were initially charged with manslaughter in this incident. Although those charges were later reduced to misdemeanors, close coordination of all agencies and individuals involved is needed if a successful conclusion to the case is to be reached. Investigating agencies must coordinate information gathering, investigation, and obtaining and safeguarding physical evidence. Generally, the fire departments should have prime responsibility for cause determination and gathering of evidence to support a determination of an incendiary fire or set fire.

4. The local alarm systems in the dormitories need to be inspected and/or tested several times during the school year, especially to check for vandalism.

5. Students need to be informed of the proper number to call when a fire is seen – usually 911. They also need to be informed of the types of equipment and practices that are outlined in dormitory rooms.

6. Sprinkler systems should be retrofit in multiple occupancy dwellings, whenever possible. Smoke detectors should be provided in every room and in hallways.

APPENDICES

Wesley College, Dover, Delaware Fire

A. Dover Fire Department Incident Report, including dormitory layout and apparatus deployment diagrams.

B. Wesley College's campus police incident report.

C. Wesley College Fire Safety Regulations and Procedures.

D. Slides of Wesley College fire, with sketches of where #1 - #16 were taken (with USFA master file copy only).

E. Pictures of Wesley College fire.

BUREAU OF FIRE
DOVER DE.
FIRE REPORT

INCIDENT NO. 155

BOX NO.

Year	Month	Day	Time		Code
1987	APR	12	0238 33		10-8
1987	APR	12	0238 —		10-8
1987	APR	12	0240 40		10-2
1987	APR	12	0454 00		10-2
1987	APR	12	0545		10-1
1987	APR	12	0547		10-19
					10-7

TYPE OF FIRE Building

UTILITIES NOTIFIED

FIRE MARSHALL CAR-3

AMBULANCE MEDIC 46 / A-43

CREW CALL

OFFICERS CALL

SQUAD 1 — W. HUTCHISON
LADDER 1 — 16-20
ENGINE 2 —
ENG LADDER 3 — E. BAKER
ENGINE 4 — J. HURD
TENDER 5 — D. LOUIC
ENGINE 6 — 17
ENGINE 7 — 19
ENGINE 8 —
BRUSH 9 — E-SAPP
CAR 15 10 — 15
DISPATCHER — Robert C. Parkento

REPORTED BY — E. GRIFFITH (PROFESSIONAL BUREAU)
LOCATION — Williams Hall Wesley College
PHONE NO. — 678-0305

SIMPLEX S-216092

STA 41
DISP 0253
10-2 0313
10-19 04:30

A-43
DISP 03:06
10-2 03:19

10-2 03:22
DISP 03:20

STA 51
DISP 02:59
10-2 03:14
05:45

MEDIC 46
DISP 02:34
10-2 02:49

2ND ALARM	1987 APR 12 0253	
3RD ALARM		
4TH ALARM		

Appendix A continued

BUREAU OF FIRE

103 SOUTH GOVERNORS AVENUE
DOVER, DELAWARE 19901

☒ General ☒ City
☐ Silent ☐ Rural
☐ Crew Call ☐ Other Dist.
☐ Officers Call

46 temp.   Weather 5 wind-mph

☒ 2nd Alarm ☐ 3rd Alarm ☐ 4th Alarm

☐ Snow/ice ☐ Rain ☒ Clear
☐ High winds ☐ T-storm

Incident No.	Mo.	Day	Yr.	Day of Wk./#	Alarm time	10-8	10-3	10-1	10-7
155	04	12	87	01-Sun.	0233	0238	0240	0400	0547

Address/Location		Apt.#	Bldg.#	Lot#
N. Bradford & Cecil Sts. Dover				

Occupant/Operators Name Williams Hall   **Telephone.**

Owner's Name Wesley College **Address** 120 N. State St. **Phone #** 736-2332

Method of alarm
☐ Fire phone - 1
☐ Alarm system - 3
☐ Radio - 4
☐ Verbal - 5
☐ Kent Center - 7
☐ Dover Police - 7
☐ Other hotline - 7
☒ Other - 9
House Phone

Nature of alarm (as reported)
☒ Structure fire - 11
☐ Vehicle fire - 13
☐ Brush fire - 12
☐ Trash/dumpster - 15
☐ Pole/wires - 45
☐ Medical assist - 32
☐ Search - 34
☐ Washdown/41 Gas Leak

☐ Service call - 52
☐ Bldg. rescue - 35
☐ Veh. rescue - 36
☐ Water rescue - 37
☐ 10-12 - 57
☐ Assist - 99
☐ False/Sys. Mal. - 71
☐ Other (explain) - 99

☒ Mutual aid received Media 46
Stas. 41-51- A-43

☒ F.M. 14 requested @ 0300 hrs.
☒ F.M. 14 on scene @ 0305 hrs.

Injuries F/F 0 Civilian 4
Deaths F/F 0 Civilian 1
Persons rescued (fire) 3
Persons rescued (other)

Action taken
☒ Extinguish - 1
☐ Rescue - 2
☐ Investigate only - 3
☐ Remove hazard - 4
☐ Stand-by - 5
☐ Salvage - 6
☐ Medical - 7
☐ No service - 9
☐ Other explain

Narrative - Explain SITUATION FOUND & ACTION TAKEN (Required on all alarms)

See Attached Narrative

Property use
☐ Public assembly - 1
☐ Educational - 2
☐ Institutional - 3
☒ Residential - 4
☐ Store/office - 5
☐ Industry/utility - 6
☐ Manufacture - 7
☐ Storage - 8
☐ Special/outdoor - 9
☐ Mobile - 9
☐ Other - 0

Construction type
☒ Fire resistant - 1
☐ Heavy timber - 2
☐ Protected steel - 3
☐ Exposed steel - 4
☐ Prtd. masonry - 5
☐ Exposed masonry - 6
☐ Brick veneer/frame
☐ Wood frame - 8
☐ Modular - 9
☐ Mobile home - 9
☐ Other - 0

Mobile Property	Year	Make	Model	Serial No.	State	Registration

Extent of damage | Fire | Smoke | Water
Object of origin | ☐ | ☐ | ☐
Room/area of origin | ☒ | ☐ | ☐
Floor of origin | ☒ | ☐ | ☐
Structure of origin | ☐ | ☐ | ☒
Extended beyond structure | ☐ | ☒ | ☐
No damage of this type | ☐ | ☐ | ☐

Detectors
☐ Present/operated - 1
☐ Present/not oper. - 3
☒ Not present - 4
☐ Not determined - 9
☐ Other - 0

Sprinklers
☐ Present/operated - 1
☐ Present/not oper. - 2
☐ Fire too small - 3
☒ Not present - 8
☐ Not determined - 9

Percent of Damage 25 % **No. of Stories** 5

Officer in charge John Willson - Fire Chief

Officer making report T.M. Whitham - Deputy
C. Boyer - Asst. Chief

Member making report N/A

Appendix A continued

			OIC	Driver(s)
Miles to scene, one way _1_ At least 1 mile; Total miles traveled _19_		Time under control _1_ H. _27_ M. Total time in ser. _3_ H. _14_ M.	Attendance, on scene _41_; Total attendance _47_	

	Mission	OIC	Driver(s)
☒ Squad-1	☐ Rescue ☒ Salvage ☐ Manpower / ☐ Air ☐ Service ☐ Other	W. Hutchison	G. Alderson
☒ Ladder-1	☒ Truck Co. ☐ Rescue ☐ Ladder pipe / ☐ Salvage ☐ Service ☐ Other	T. Whitham (16)	Frt. J. Hurley / Tlr. G. Alderson
☐ Engine-2	☐ Pump ☐ Water ☐ Wagon / ☐ Service ☐ Other		
☐ Ladder-3	☐ Service Aerial ☐ Ladder pipe / ☐ Rescue ☐ Service ☐ Other	•	
☒ Engine 3	☐ Pump ☒ Wagon ☐ Water / ☐ Service ☐ Other	E. Baker	R. Osika
☒ Engine-4	☐ Wagon ☐ Pump ☐ Water / ☒ Manpower ☐ Service ☐ Other	J. Hurd	A. Moore
☒ Tender-5	☐ Hose ☐ Master streams ☐ Rescue / ☒ Manpower ☐ Service ☐ Other	D. Louie	E. Moore
☒ Engine-6	☒ Wagon ☐ Pump ☐ Water / ☐ Service ☐ Other	C. Boyer (17)	H. Pusey
☒ Engine-7	☐ Wagon ☐ Pump ☐ Water / ☐ Service ☒ Other	W. Hurley (19)	C. Carey Sr.
☐ Engine-8	☐ Pump ☐ Hose ☐ Service / ☐ Other		
☒ Brush-9	☐ Field Piece ☐ Manpower ☒ Other		E. Sapp
☐ Boat	☐ Rescue ☐ Service ☐ Other		
☐ Foam Trl.			
☒ Car	☒ Command Post ☐ Manpower / ☐ Service ☐ Invest.	J. Willson (15)	S/A

Equipment used
☐ Air bags
☒ Air masks _18_
☒ Cascades _1 & 1_
☐ First aid eqt.
☒ Forc. entry eqt.
☒ generators _L1 & S-1_
☐ hurst tools
☐ K-12 saws
☐ ladder pipes
☐ salvage eqt.
☐ sal. covers
☒ misc. sm. tools
☐ winches
☐ Other Equip.

Hose used - Feet
Booster _0_
1½" _100_
1¾" _200_
2½" _250_
3" _50_
5" _300_
Total _900_

Master Streams Used
☐ Lad. 1   ☐ Eng. 6
☐ Lad. 3   ☐ Eng. 7
☐ Eng. 3   ☐ Eng. 8
☐ Eng. 4
☐ ☐ ☐ Tender 5

Ladders raised - Ft.
Aerial _100_
Bangor _____
Wall _90_
Extension _48_
Roof _____
Collp. _____
Step _8_
Step to Straight _8_
Total Ft. _____

_____ Salvage covers left @ scene
_____ Port. Pumps left @ scene

Gals. of water _1,500_
Gals. of Hi-X _0_
Gals. of 6% _0_
Gals. of AFFF _0_
Gals. of Protein _0_
No. of Hydrants _2_
Total Tank Water _0_

Extinguisher Used
☐ Co²   ☐ Dry Powder
☐ Halon   ☐ Metal-x

Officers present ☒15 ☒16 ☒17 ☐18 ☒19 ☒20

Police officers at scene C.I. Capt. Hutchinson, Cpl. Stallings   ☒ Dover P.D. ☐ Capitol P.D. ☐ Del. State

Utilities at scene

Misc. officials at scene College Security, College Maintenance, Dr. Reed Stewart

Alarm taken by Robert C. Tarburton

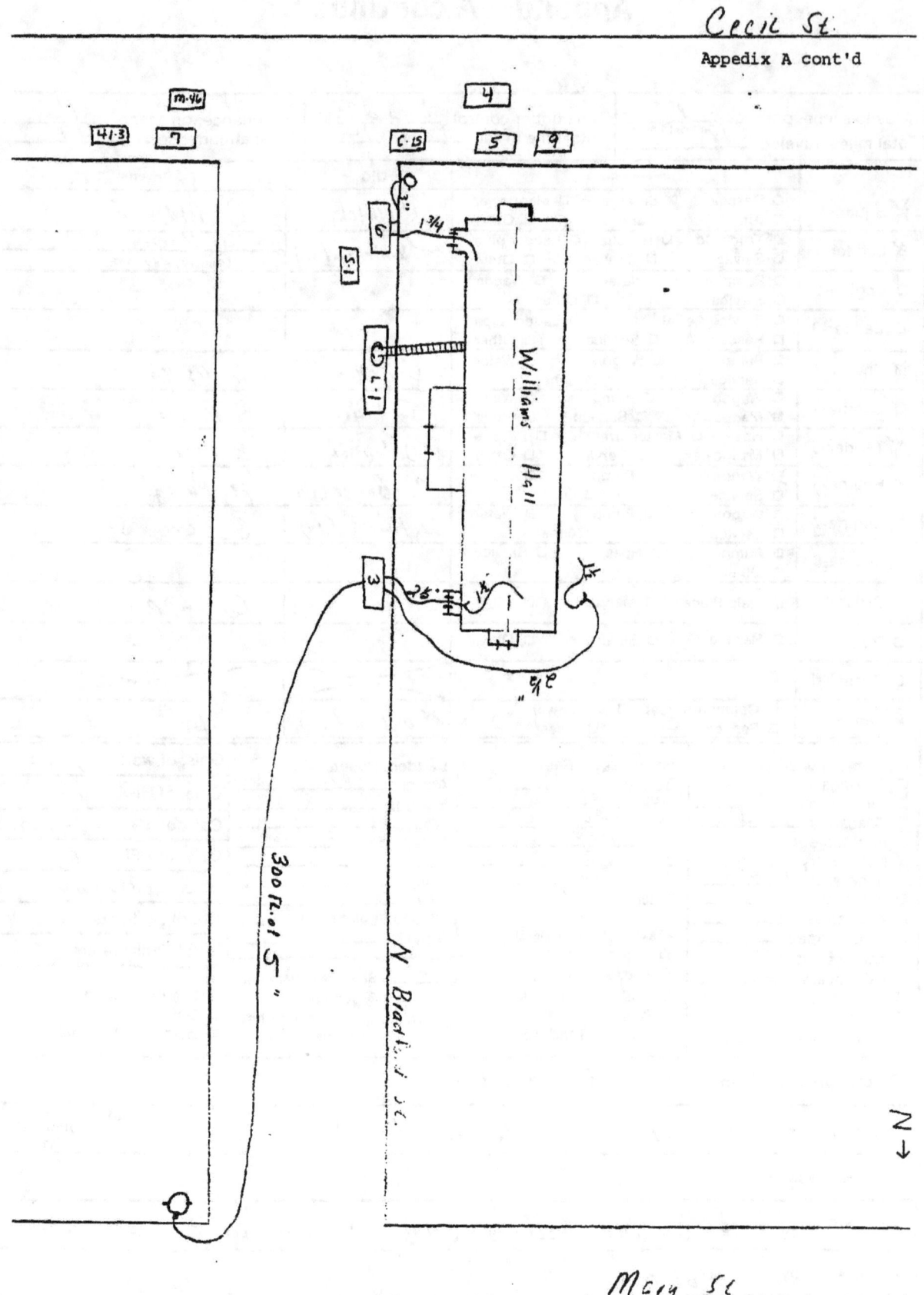

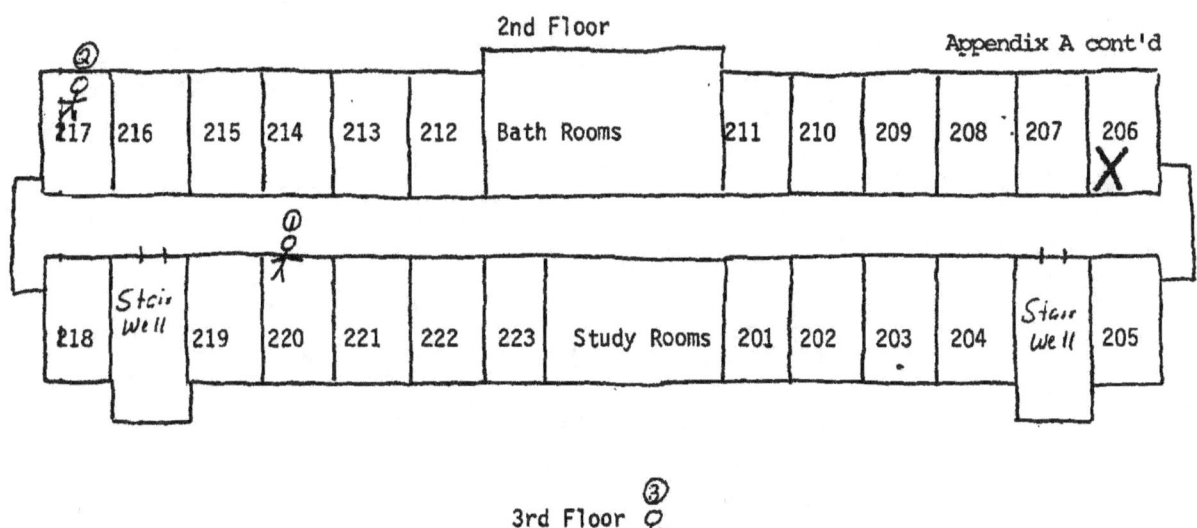

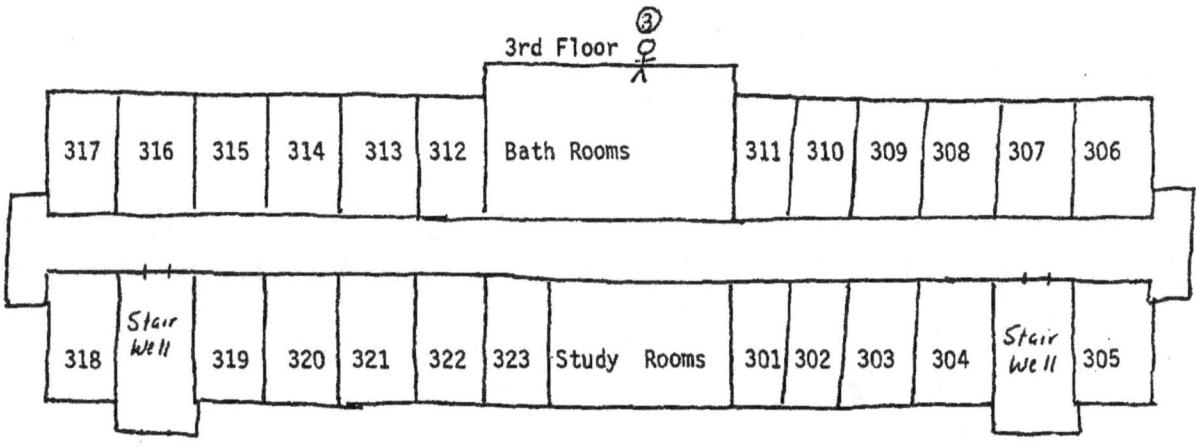

X - Fire Origin

1.) D.O.A.

2.) Unconscious Victum

3.) Unk. male rescued by F.D. ladder from rear 3rd floor window

Williams Hall floor plan (2nd & 3rd floors)
Wesley College

N →

APPENDIX B

WESLEY COLLEGE
OFFICE OF THE DEAN OF STUDENTS
STAFF REPORT

FROM: Randy Johnson   TO: Dr. J. Thomas Sturgis

HALL: Williams Hall   DATE: April 12, 1987

RE: Fire Alarms

At 1:20 am, Saturday, April 11, 1987 the fire alarm sounded. Steve MacGlashan and myself left the office area and headed to the third floor. I stopped at the second floor to check that area. There were people in the hallway but no one seemed to be going to the exits. There was no sign of fire or smoke, however, the alarm was pulled on this floor. Steve and myself proceeded, to the third floor (which also showed no signs of fire) and began knocking on each door with the purpose of clearing the building. After clearing the third floor, we began the same process on the second floor. During this period, the security guard (Russell Pleasanton) asked for my permission to turn off the alarm. I denied his request until the building-was completely checked. At that time, I told the security guard to turn off and reset the alarm and he proceeded to do so. Approximately 1:40 am the process was completed. At 3:30 am, I was informed by Russell (security guard) that a bell had been stolen from the north end of the first floor hallway.

REFERRED TO: _____ DATE: _____

R186

Appendix B continued

WESLEY COLLEGE INCIDENT REPORT

Location of Incident	WILLIAMS HALL 2nd FLOOR NORTH END.	Time of Incident	01:20
		Time of Discovery	

Description of Incident: (Including Seriousness of Property Damage, College Regulations Violated, Extent of Personal Injury)

FIRE ALARM PULLED IN WILLIAMS HALL, 2nd FLOOR NORTH END.
HELPED RD AND RA's TO EVACUATE BUILDING. AFTER ALL WERE
EVACUATED. FIRE ALARM ON 2nd FLOOR NORTH END WAS
RESET AT 01:30 HOURS.

Persons Involved: (Witnesses, Offending Parties, Offended Parties)

SECURITY

Person (s) Notified of Incident:

RD, SECURITY

Report Filed By: Russell D. _____

Date: 11 APR 87

Appendix B continued

TAMPERING WITH FIRE ALARM, FIRE EXTINGUISHERS AND RELATED FIRE EQUIPMENT WILL NOT BE TOLERATED AND WILL RESULT IN IMMEDIATE COLLEGE DISCIPLINARY ACTION, AS SOON AS THE CITY FIRE MARSHALL IS NOTIFIED THE FALSE ALARM IMMEDIATELY BECOMES EITHER A CITY OR STATE OFFENSE OR BOTH, RESULTS OF THE CITY FIRE MARSHALL'S INVESTIGATION MAY RESULT IN CRIMINAL PROSECUTION OF THE INDIVIDUAL(S) INVOLVED IN ADDITION TO ANY COLLEGE DISCIPLINARY ACTION

Fire Evacudtion Route

FIRST FLOOR

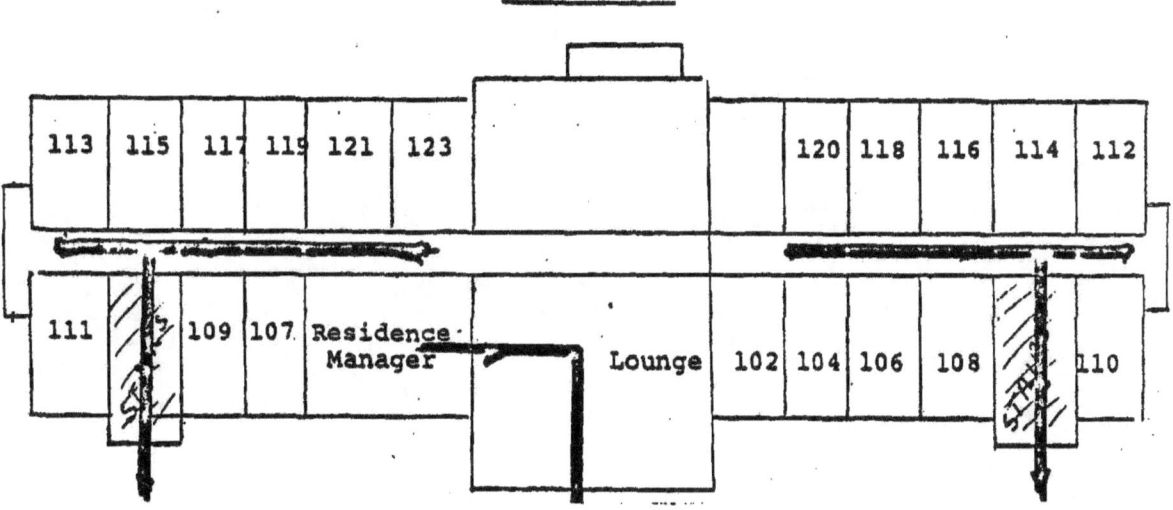

APPENDIX C

Wesley College
Fire Safety Regulations and Procedures

A. Fire Safety Regulations

1. Smoking is not permitted in specific areas on the campus.

2. Smoking in bed is not permitted.

3. Hot plates, coffee pots, and toasters are not permitted in residence hall rooms

4. Excessive use of extension cords is prohibited.

5. Only approved electric lighting is permitted in student rooms

6. Stairway fire doors must remain closed at all time.

7. Residents are responsible for maintaining clean rooms at all times.

B. Fire Alarms and Fire Drills

1. Any time a fire alarm sounds in a residence hall, it must be responded to promptly and efficiently. It is imperative that all persons react as if there is a fire. Never assume it is a false alarm. Prompt responsible action will result in minimizing or preventing injury to persons in the building and damage to equipment.

2. Fire alarms will be tested monthly by college officials. Students must evacuate building during testing unless advance notification of the test is posted.

3. Fire drills will be conducted at specified times. Students are required to evacuate the hall during these drills. Students must respond in the same responsible manner they would for an actual fire.

4. Tampering with fire alarms, fire extinguishers, and related fire equipment will not be tolerated and will result in immediate college disciplinary action.

C. Procedures in Event of Fire or Fire Drill

1. The first person discovering the fire should pull the nearest alarm and immediately report to the Residence Director.

2. All students should immediately evacuate building in a rapid, calm, and orderly manner. If possible, each student should ensure that their room door is open, lights on, windows closed, and blinds raised.

19

3. After leaving the building, students should report to the following areas and remain there until given additional instructions by the Residence Director and/or other college or city fire officials. Students may not re-enter the building until permission is granted from the college official in charge..

    Carpenter Hall – Parking lot behind Dulany Hall.

    Williams Hall – Slaybaugh Hall lawn across street.

    Budd Hall – Parking lot across from Budd Hall.

    Cecil Hall – Tennis courts behind gym.

    Gooding Hall - Cannon Hall lawn across street.

    (In event of inclement weather, other areas of assembly will be designated.)

4. Resident assistants should be the last person to leave the floor and should verify that all students have left by checking rooms and closing doors behind them. Resident assistants should then report to the assembly area, account for all persons on their floor, and report to the Residence Director.

5. Resident assistants should immediately check the fire boxes on their floor to ascertain if the alarm has been pulled. If it has, report this immediately to the Residence Director.

6. The Residence Directors, immediately upon hearing the alarm, should notify college security for assistance. The security guard should report immediately to the Residence Director.

7. The Residence Director will not call the fire company until there is reasonable indication of an active fire, i.e., smoke, report by another person of fire to the Residence Director, personal observation of fire, etc. If, after three minutes the Residence Director has not received a direct indication of fire, an inspection of the building should be conducted.

8. Immediately upon reasonable determination of fire, the Residence Director will notify the fire department giving building name, address, location of the fire, type of the fire, and name of the person calling. Then, she will notify the Dean of Students.

Slides of Williams Hall
Wesley College, Dover, Delaware

The numbers are those that appear on the enclosed selected slides. The sketches following show where slides 1-16 were taken.

1-3. Front views of Williams Hall. Slide #1 shows scorching outside the room of origin.

4. Identification plaque on front of Williams Hall.

5. Alarm bell and pull station in basement hallway.

6. Close-up of pull station in basement hallway.

7. Top of stairs on fire floor looking toward corridor south end. Note heat damage on top half of doors at both ends of the corridor (slides 7 & 8).

8. Top of stairs on fire floor looking toward corridor north end.

9-14. Room of origin (room #206).

13. A close-up of the point of origin and the warped door frame due to intense heat.

14. Note the similarity of the area of origin with the area on the wall where a cabinet was.

15. Corridor facing south toward room of origin.

16. Burned wall and ceiling fixtures 12 inches distant from room of origin.

23. Undetermined, similar to room of origin. Injured student in room 217 was found under the desk area.

26. Missing fire alarm bell on first floor opposite room #108.

27. Fire extinguisher not used.

28. First floor corridor.

29. Resident assistant office area located at main entrance to dormitory.

30. Close-up missing alarm and bell, first floor.

31. Stairwell door opposite room #220.

32. Student room door with self locking door knob.

33. X-rated message on ceiling of third floor. Appears to be written using candle. This is an example of similar "slogans" throughout the dormitory on walls and ceilings, using candles and lighter fluid.

34. Close up of wall light with apparent burn marks from vandalism.

35. Examples of complete disregard for fire safety found throughout the building.

36. Second floor window outside room of origin.

APPENDIX E

Pictures of Wesley College Fire

1. Front of Williams Hall; deceptively light looking smoke damage is seen above room of origin window at right on second level.

2. Damage in corner of room of origin. Note that room (in right of photo) across the hallway from the room of origin is relatively undamaged because its door was closed.

3. Hallway damage immediately outside the room of origin. Ceiling damage is also visible. The fire was extremely hot.

4. Long shot down hallway outside the room of origin. The student who was killed was in a room at the far end of the hallway.

5. The alarm and bell were missing on the first floor at the time of the fire.

6. Light fixture in a hallway on another floor of the dorm had been damaged by an open flame; there were several signs of previous minor vandalism caused by open flames.

Photo 1. Front of Williams Hall; deceptively light looking smoke damage is seen above room of origin window at right on second level.

Photo 2. Damage in corner of room of origin. Note that room (in right of photo) across the hallway from the room of origin is relatively undamaged because its door was closed.

Photo 3. Hallway damage immediately outside the room of origin. Ceiling damage is also visible.
The fire was extremely hot.

Photo 4. Long shot down hallway outside the room of origin. The student who was killed was in a room at the far end of the hallway.

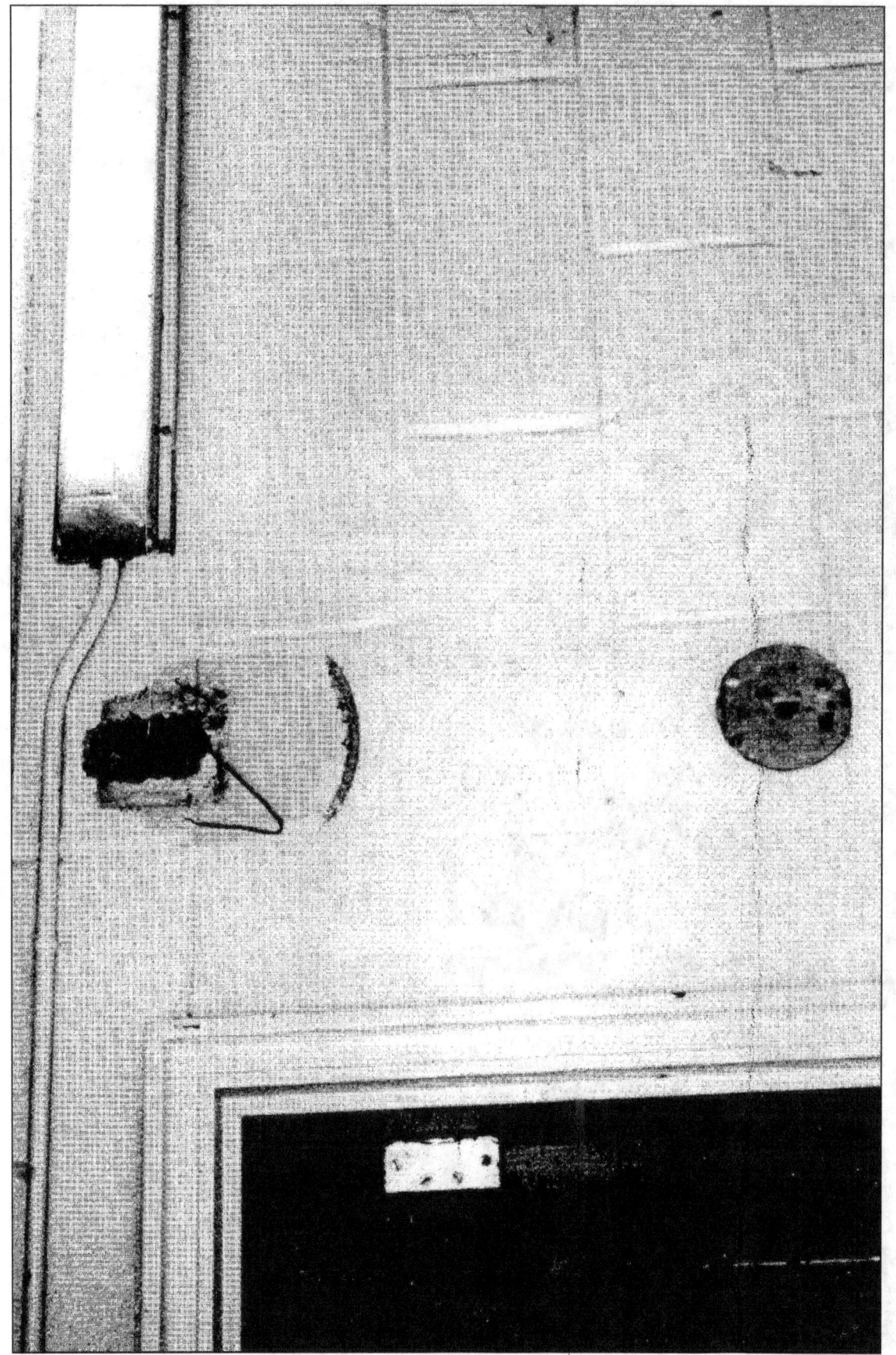

Photo 5. The alarm and bell were missing on the first floor at the time of the fire.

Photo 6. Light fixture in a hallway on another floor of the dorm had been damaged by an open flame; there were several signs of previous minor vandalism caused by open flames.

LONGWOOD COLLEGE DORMITORY FIRE
Farmville, Virginia, April 28, 1987

Local Contacts: **Virginia State Fire Marshal's Office:**
Daniel Alterescu
Deputy Fire Marshal
Jim Reed
Fire Protection Engineer
205 N. 4th Street
Richmond, VA 23219
(703) 786-4751

Longwood College, Farmville, Virginia:
Richard V. Hurley, VP Bus. Affairs (804) 392-9225
Sue A. Saunders, Ph.D., Dean of Students (804) 392-9221
Richard A. Weibl, Director of Housing (804) 392-9233
Randy L. Dean, Resident Education Coordinator (804) 392-9374
Sgt. J. A. Huskey, Campus Police Investigator (804) 392-9321, -9223
Melvin Moore, Physical Plant Supvr.
Bill Brown, Prev. Maintenance Supvr.

INTRODUCTION

On the morning of Tuesday, April 28, 1987, at approximately 6:50 a.m., a fire started on the third floor of the Frazer Dormitory on the campus of Longwood College in Farmville, Virginia and caused fifteen students to be treated for injuries. Of these, 12 were treated for smoke inhalation, one for second- and third-degree burns, one for a broken ankle, and one for severe respiratory problems partially caused by a previous illness.

The information contained in this report is based on personal observations and interviews with the persons whose names appear above. In addition, bits and pieces were gathered from others who were at the scene during the fire. Farmville Volunteer Fire Department Chief Philip Gay and Virginia State Police Investigating Officer Roy Kyle were not available for interview during this investigation, and no incident reports were available from the fire department.

THE FIRE

On the morning of the fire, at approximately 7 a.m., "heavy smoke and orange flames" were observed by Morris Walker, boiler plant employee, coming from the third floor corner window of Frazer Dormitory. Subsequently, several telephone calls were received by the Campus Police dispatcher, who, in turn, called the Farmville Police and the Farmville Volunteer Fire Department.

According to several members of the faculty, the fire was first observed in the room of origin by a student occupant of the room, Garland Barr, who was awakened up by the fire. Mr. Barr stated that the stereo speaker appeared to burst into flames. While he was trying to put the "stereo fire" out, the fire spread to tie-dyed sheets which had been hung in the middle of the room for decoration and to provide the occupants with additional privacy. Deputy State Fire Marshal Jim Reed stated that "the probable cause of the fire was an overloaded light-duty 8-foot long, 6-outlet electrical extension cord, which was not approved." The stereo was plugged into that cord. In addition, it appeared that knots had been tied in the cord for no apparent reason, but it was not exactly clear how and why the fire started.

As the fire continued to spread, the dormitory hall fire alarm was pulled by a student but failed to operate. It was later determined that the alarm system did not activate because the main breaker switch for the alarm system (in the basement) was in the "off" position. It was finally activated manually by a dormitory resident assistant approximately 10 minutes after the fire was discovered. In addition, a follow-up inspection revealed that approximately 85 percent of the smoke detectors in student room were either disconnected or failed to operate. The detector in the room of origin also apparently did not work.

Students who became aware of the fire began knocking on doors to alert other students, who thought it was "just another fire drill." Several students ignored even the vocal alarms until an announcement was made over the P.A. system that "this is not a joke, this is a real fire."

As students exited the third floor down the stairs, large quantities of smoke and hot gasses which had been accumulating on the third floor entered the stairwell. Several existing students suffered smoke inhalation.

The campus police who were first notified and first to arrive had no firefighting protective gear, no breathing apparatus, no firefighting training. They could not fight the fire or undertake rescue in the situation found.

Upon arrival of the first fire department unit, students were still exiting the building. The first arriving firefighters delayed their entry into the building to assist in the evacuation. Entry to the fire floor was not made immediately because the response to another fire shortly before had depleted their supply of air pacs and severely limited breathing equipment. Upon arrival of the aerial ladder truck (purchased by Longwood College for use of the fire department to reach high- rise buildings), its breathing equipment with full air pacs was used and entry made.

The fire was controlled from spreading to upper floors by an outside attack using the aerial ladder and handlines stretched via ground ladders up to the canopy over the entrance. Only minimal interior firefighting was apparently accomplished, except for extinguishing spot fires during overhaul.

Structural damage was primarily limited to the room of origin. The fire resistant construction and firefighting operations were apparently sufficient to curtail flame spread. No dollar estimate of property loss is available at this time.

THE CONSTRUCTION

Frazer Dormitory, is a ten-story, fire resistant building with a basement, brick exterior, and concrete block interior. It was built in 1970 before many of the fire protection requirements of today were mandatory. There were no sprinklers.

Doors to the student rooms were not self-closing. Locks varied in type and could not be opened by a master key.

As a result of the fire, several recommendations were to be made by the Virginia State Fire Marshal's Office to improve the building. However, the most recent fire code was adopted in 1984, and strict adherence to current codes cannot be mandated by current Virginia State law.

LESSONS LEARNED AND RECOMMENDATIONS

As in many schools, the basic Longwood College safety policies appeared to be sound, but the process needed to insure compliance was fragmented and in need of improvement. The problem is not a lack of concern on the part of the school or its employees, but rather a lack of knowledge or authority to carry out the functions needed to obtain compliance by students and faculty.

Members of the faculty and staff at Longwood were exceptionally cooperative with this investigation. Their realization of the potential disaster which could have resulted at the college has made them keenly aware of the circumstances surrounding the fire and the need for improvements.

Lessons learned and recommendations are listed below. The intent here is to help other schools and fire officials also learn from this fire.

1. Because of the high incidence of false alarms in dormitories, many schools such as Longwood have the policy that someone – often a resident assistant – is to check on whether an "emergency" is real before calling authorities. Often the first authority called for emergencies is the campus police. These actions can delay contacting the fire department when there really is a fire, as happened in these two fires.

    When an actual fire is detected the first call should be to the 911 emergency number. Campus police and other agencies may then be notified as required. (911 alert cards have been recommended to be placed at each telephone in the Longwood dormitory.)

    According to Longwood Campus Police Chief Jim Huskey, the campus police department at Longwood is normally alerted by the Residence Education Coordinator or duty resident assistant of fires or other emergencies on campus. The dispatcher notifies the officer on-duty who then responds to the location of the emergency. Upon their arrival, they take whatever action they deem necessary to control the situation. This includes deciding whether or not to call the fire department. This is satisfactory unless a real fire is known to exist. The campus police, like industrial fire brigades, may be the first to arrive at the scene of a fire on campus, but the fire department should also be called.

2. Where campus police are to be the first responding agency called to control a fire situation, as at Longwood, they need to be adequately staffed and properly trained and equipped to enter a burning, smoke-filled building. Protective clothing, breathing equipment, and fire extinguishers need to be kept in their vehicles. They also must be trained on when and under what circumstances the fire department will be called.

3. The campus police depend on students to complement their numbers. A sufficient number of auxiliary student police should be recruited to insure adequate first response manpower at all times. Basic fire brigade training should be given at the beginning of each semester to campus police officers and the auxiliary student police.

4. Fire safety training for Residential Education Coordinators, Residential Assistants, and students should be given at the beginning of each school year. Follow-up training would also be beneficial. This training should include instructions on the use of the 911 emergency number.

5. Doors were not self-closing. They should be. This could have helped contain fire and smoke in the room of origin.

6. Access to student rooms must be made quickly in an emergency to see if they are occupied and if anyone is injured or needs assistance to evacuate. The variety of individual door locks at Longwood necessitated a large ring of keys to be maintained and locked in the first floor Resident Education Coordinator's office. Consequently, the keys were not used to open any of the students' rooms to ascertain that all rooms had been evacuated. The door locks should be changed, perhaps during summer break, and a lock box kept on each floor for access by that floor's residence assistant during an emergency.

7. Bed linens and other flimsy combustible materials were the probable cause of the rapid spread of this fire. Hanging bed linens as room dividers appeared to be a frequent practice throughout the dormitory. This should be forbidden. Occasional unannounced room inspections might be used to check on compliance and other safety practices. The forbidden practices and equipment (e.g., the 6-outlet extension cord) should be stressed in fire safety training and literature provided to the students.

8. Exit doors leading to the roof were locked at the time the fire occurred, thereby preventing egress. Panic hardware with an automatic alarm would allow egress in an emergency but prevent unauthorized personnel from using this exit. Breakaway locks and/or keys should be readily available.

9. Firehose and cabinets adjacent to the riser were removed after the fire, because the 1 1/2 inch firehose was not equipped with a 2 1/2 inch adapter. These should be replaced, equipped with compatible hardware, and locked.

10. Access to the front of Frazer Hall by ladders was obstructed by overhanging porches and a low electrical power line. The lack of a driveway at the rear of the building could also impede the arrival of fire apparatus attempting to gain access from that side of the building. Such features should be considered in design reviews of new buildings. Pre-fire plans should be made to cope with them if they already exist. They can also be part of the argument to retrofit sprinklers in such high-rise residential structures.

11. Corridor walls in the dormitory were breached with 2 inch to 3 inch holes to allow for a TV cable to be put in each room. No attempt to plug these holes had been made. Such breaches should be corrected immediately using appropriate sealing methods. They can allow smoke and flames to spread.

12. Buildings and facilities of this type should receive at least annual inspection by the State Fire Marshal's Office and more frequently by college staff. (Longwood dormitory was scheduled for inspections once every three years.)

13. Ventilation of buildings during a fire should be accomplished by fire department personnel only. This operation must be done systematically to prevent additional smoke damage and possible spread of the fire. During this fire, it was accomplished by one of the Physical Plant personnel.

14. Sprinkler systems could have extinguished this fire before it caused any injuries. They should be installed in all residences.

15. Smoke detectors should be regularly tested and maintained in working order. Dormitory alarm systems also must be regularly checked, especially to ensure they have not been vandalized.

The most difficult task to implement is changing attitudes of students and school officials to take the threat of fire seriously. For at least a while, Longwood College will have the proper motivation. Other colleges and universities should consider these lessons for themselves. These situations are common to many.

APPENDICES

Longwood College, Farmville, Virginia

A. Longwood College Campus Police Incident Report and Farmville Fire Department Incident Report.

B. Frazer Hall – Residence Hall Fire Safety Procedures.

C. Pictures of Longwood College fire.

D. Slides of Longwood College fire, with sketches of where slides were taken.

E. Xeroxes of additional fire photos (USFA master file copy only).

F. Report from Longwood College Internal Investigation on status of fire alarms before the fire (file copy only).

APPENDIX A

ABSTRACT OF POLICE REPORT LONGWOOD CAMPUS POLICE
FARMVILLE, VIRGINIA 23901

COMPLAINT NUMBER: __870428021_ OFFENSE: ____fire_____

COMPLAINANT/VICTIM: _____Morris Walker_____

ADDRESS: _Longwood employee-boiler plant_____

LOCATION OF INCIDENT: _____Frazer_____

TIME OFFENSE OCCURED: __0710_____ DATE: ___4/28/87_____

ITEMS REPORTED STOLEN (IF APPLICABLE):

_____ VALUE: _____

_____ VALUE: _____

_____ VALUE: _____

SUSPECT: _____ STUDENT? _____

SUSPECT: _____ STUDENT? _____

SUSPECT: _____ STUDENT? _____

BRIEF NARRATIVE:
At 0710 we received a telephone call from complainant. He stated
that he could see fire and smoke coming out of Frazer Dorm. When
we arrived we found fire and smoke coming out the corner room
facing Spruce Street on the East end of the building (third
floor). The alarm was going off and students were coming out of
the building. We called Farmville Fire Dept. and was told by the
dispatcher that he was aware of the fire and the fire trucks were
on the way. Officer Campbell stayed in the street to work
traffic. Officer Sudesberry and I attempted to get to the rooms
on the third floor, but was unable to do so because of the smoke.
the fire trucks arrived at this time.

WAS ALCOHOL A FACTOR? YES _____ NO _____
POTENTIALLY PREVENTABLE? (FROM OFFENSE REPORT) YES_____ NO_____

                    I CERTIFY THIS IS A TRUE ABSTRACT OF THE
                    INITIAL REPORT FROM OUR OFFICIAL POLICE
                    RECORDS.

DATE: 04/29/87   _____Officer Nunnally_____
                             CERTIFYING OFFICER

39

Appendix A continued

COMPLAINT NUMBER				TYPE OF OFFENSE			
Yr 87	Mo 04	Day 28	No. 021	Fire			
OFFENSE CHANGED TO AFTER INVESTIGATION:					TIME OF REPORT 0900	DATE OF REPORT 4-28-87	

NAME OF VICTIM N/A	VICTIM'S WORK/STUDENT ADDRESS N/A	

VICTIM'S HOME ADDRESS N/A	VICTIM'S PHONE N/A	COMPLAINANT'S PHONE 392-6671 —
	Office Home	Office Home

COMPLAINANT'S NAME Morris Walker (employee-boiler plant)	COMPLAINANT'S ADDRESS Boiler plant

LOCATION OF INCIDENT Frazer	TIME OFFENSE OCCURRED FROM 0710 TO —		

INVESTIGATOR ASSIGNED	OFFICER ASSIGNED 302-206-210	PERSONS ARRESTED N/A	STUDENTS ☐ YES ☑ NO	EMPLOYEE ☐ YES ☐ NO

At 0710, we recieved a telephone call from complainant. He stated that he could see fire and smoke coming out of Frazer dorm. When we arrived we found fire and smoke coming out of the corner room facing Spruce St on the east end of the building (third floor)

The alarm was going off and students were coming out of the building. We called F.P.D. and was told by the dispatcher that he was aware of the fire and the fire trucks were on the way. Officer Campbell stayed in the street to work traffic. Officer Sudsberry and I attempted to get to the rooms on the 3rd floor, but was unable to do so because of the smoke. The fire trucks arrived at this time.

WANTED PERSONS

Attach Lookout Cards	N/A

WANTED PROPERTY N/A

PROPERTY STOLEN N/A

STATE PROPERTY ☐ YES ☐ NO	ALCOHOL A FACTOR ( ) YES ( ) NO N/A
	POTENTIALLY PREVENTABLE ☐ YES ☐ NO N/A

ITEM DESCRIPTION	VALUE
N/A	N/A

REPORTING OFFICER Nunnally	DATE 4-28-87	APPROVING SUPERVISOR	DATE

Appendix A continued

FARMVILLE VOLUNTEER FIRE DEPARTMENT
FIRE REPORT

Attendance	P/V	
1. _Witt_		
2. _T. Coile_		
3. _Buddy Coile_		
4. _Phillip Bailey_		
5. _Mark Mills_		
6. _Billy Johnson_		✓
7. _Jimmy Price_ 34		
8. _W. Moss Jr_		
9. _Robert Spruel_		
10. _Al Mason_		
11. _Palmer Saunders SR_		
12. _T.J. Ester_		
13. _E. Ross_		
14. _Chris Warner_		✓
15. _John Ague_		
16. _Gary Atkins_		
17. _Bobby Ward_		
18. _Chief Gay_		
19. _Cobb_		
20. _Ronnie Richardson_		
21. _John T. Coile_		✓
22.		
23.		
24.		
25.		
26.		
27.		
28.		
29.		
30.		
31.		
32.		
33.		
34.		
35.		
36.		
37		
38		
39		
40.		

Date _4-28-87_

Time of Alarm _7:00_

	31	32	33	34	35	36	38
Time Out	7:10	7:06	7:07	7:00			7.04
Time On Scene							
Time Out Of Service	9:32		9:33	9:25		9:32	9.30

(37)-9:23

Miles Travelled _____

Exact Location _LONG WOOD COLLAGE_

Type of Incident or Fire _____ Brush or Field ✓ Structure _____ Other

(Explain) _____

Type of Structure _Brick_

Cause of Incident or Fire _Under investigation by State_

Occupant _Students_

Address _____ Telephone No. _____

Owner _STATE. VA_

Value of Property _____

Estimate of Damage _____

Vehicle _____ Year _____ Make _____ Model _____

_____ License Number

Officer Answering Alarm _Chief Gay_

Officer in Charge _Chief Gay_

Apparatus Used _#31 #32 #34 #38 #33_

Equipment Used _Water light. Axe_

Mutual Aid Required _Yes_ Assisted by _Hampden Sydney & Prospect_

Description of Incident (action taken to control)

_____
_____
_____
_____
_____
_____
_____

Frazer Hall – Residence Hall Fire Safety Procedures

(The procedures below are printed on a bright orange, 4 by 6-inch sticker.)

RESIDENCE HALL FIRE SAFETY PROCEDURES

1. When a fire alarm sounds you Must evacuate the residence hall. Failure to evacuate may be hazardous to your health and result in disciplinary action.

2. Check the door of your room/suite:

    a. If cool, proceed with evacuation.

    b. If warm, open door slowly to see if it is safe to exit.

    c. If hot, DO NOT open. Stand near window and wait for help.

3. Exit room, closing but not locking the door. Take your keys with you. Always wear shoes and protective clothing.

4. Go to closest exit, enter stairwell, and leave the building as quickly as possible. DO NOT attempt to use the elevator.

5. Re-enter the building only when directed to do so by either a residence hall staff member or campus police officer.

Tampering with fire safety equipment, alarms, smoke detectors, and this notice is a violation of college policy and State law.

APPENDIX C

Pictures of Longwood College Fire

1. First handline on the fire. Note campus police with no protective clothing.

2,3. Ladder truck in action.

4,5. Room of origin.

6. Nearby dormitory room for comparison.

Photo 1. First handline on the fire. Note campus police with no protective clothing.

Photo 2. Ladder truck in action.

Photo 3. Ladder truck in action.

Photo 4. Room of origin.

Photo 5. Room of origin.

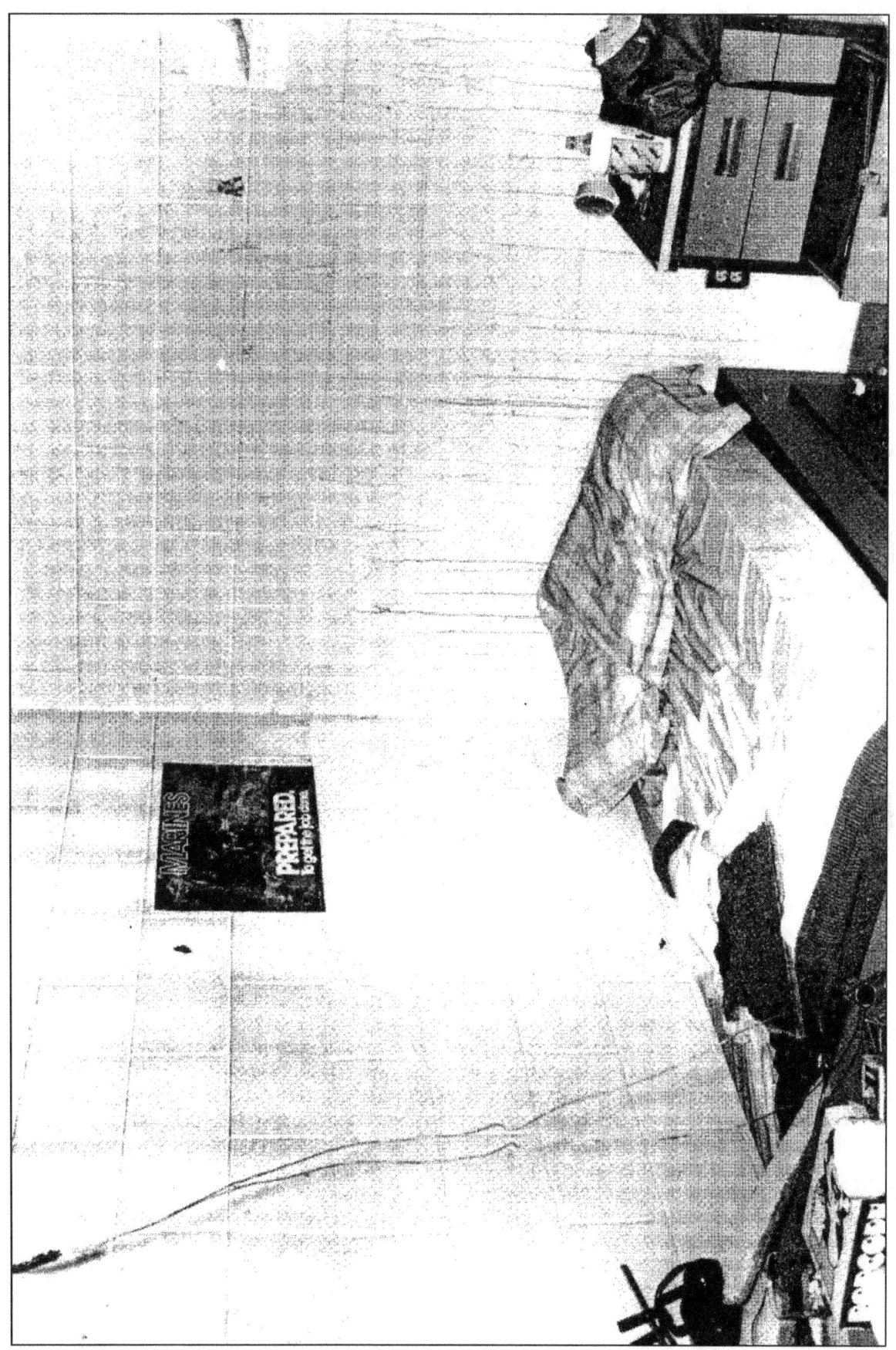

Photo 6. Nearby dormitory room for comparison.

www.ingramcontent.com/pod-product-compliance
Lightning Source LLC
Chambersburg PA
CBHW081226170526
45165CB00009B/2977